Anti-Matter and Black Holes

Anti-Matter and Black Holes

Adam G. Freeman

iUniverse, Inc.
New York Lincoln Shanghai

Anti-Matter and Black Holes

iUniverse books may be ordered through booksellers or by contacting:

iUniverse
2021 Pine Lake Road, Suite 100
Lincoln, NE 68512
www.iuniverse.com
1-800-Authors (1-800-288-4677)

The views expressed in this work are solely those of the author and do not necessarily reflect the views of the publisher, and the publisher hereby disclaims any responsibility for them.

ISBN-13: 978-0-595-43093-2 (pbk)
ISBN-13: 978-0-595-87433-0 (ebk)
ISBN-10: 0-595-43093-7 (pbk)
ISBN-10: 0-595-87433-9 (ebk)

Printed in the United States of America

Anti-Matter and Black Holes[1]

Prevailing views in physics towards anti-matter and black holes are incomplete and inaccurate because the geometry of anti-matter and black holes is misunderstood. By giving a clear and complete ontological and mathematical solution to anti-matter and black holes and describing the correct geometry, this book improves upon existing physics understanding and research. Anti-matter is shown to intrinsically have a negative spacetime curvature, to exist in a 'spacelike' Minkowski spacetime as opposed to a 'timelike' or 'lightlike' Minkowski spacetime, to move backwards in time with a proper time $-\tau$ and to have negative energy $-E$. Also, by solving mathematically for the 'mysterious parameter β' elicited from David Hestenes' real Dirac equation, it is revealed that β differentiates between the 'spacelike', 'timelike' and 'lightlike' states of a particle and in fact describes what happens at and past the event horizon in a black hole as a particle traverses through it. This reinterpretation of anti-matter and black holes is referred to as the '*Theory of Compressed Spacetime for Anti-matter*' or CSA for short.

Contents

I.   Introduction 1

II.   Theoretical Basis for the Theory 3

III.   The Anti-Particle in a Box 13

IV.   The Unexpected Lorentz Transformation 18

V.   'Spacelike' Minkowski Spacetime 21

VI.   The Real Dirac Equation 25

        The Mysterious Parameter 'β' 30

VII.   The Motion of a Particle through a Black Hole 34

VIII.   The Freeman Drive 37

IX.   Mathematical Form of β 39

X.   Backwards in Time? 42

XI.   The Anti-Hydrogen Atom 44

XII.   General Relativity 45

XIII.   Conclusion 46

XIV.   References 48

I.

Introduction

When Schwarzschild discovered black holes as being a consequence of general relativity, he made certain assumptions about the geometry of a black hole. His discovery was based on a black hole having a spherical geometry and a real radius r. The density of a black hole was found to be so massive that at a certain radius, the Schwarzschild radius, not even light waves could escape the gravitational pull of the black hole. To this day, this incorrect conception of what a black hole is has remained the same. Other physical theories involve a black hole having thermodynamic properties; absorbing, altering and re-emitting particles through quantum perturbations just outside of the Schwarzschild radius and dissipating energy and even evaporating over time.

The issue with all of these physical theories is that they are predicated on the inaccurate conception and geometry of a black hole. Conceptually, a black hole is perceived as a big cosmological dumping ground that no particle can escape from. In fact, other than skipping around the issue with quantum perturbation and virtual pair production, no particle ever reaches the event horizon of the black hole. A particle becomes successively closer and closer to the event horizon but never in fact reaches it. If the particle is not of quantum proportions, then the thermodynamic properties of black holes do not apply. So if

1

the black hole sucks in a planetary system, that planetary system will not be 'altered and re-emitted' since the planetary system will never get to the event horizon and will never actually enter the black hole. This does not make sense and this book shows otherwise. In fact, the thermodynamic solutions for the black hole including Hawking radiation are incorrect because they solve for a black hole that is spherical and a black hole is not spherical.

This book casts a new light on what a black hole really is and mathematically reveals its true nature. A black hole has a pseudo-spherically symmetrical and hyperbolic geometry, a negative spacetime curvature and the Schwarzschild radius is in fact an imaginary ir. There are phenomena in physics that have the same geometry, the anti-particles and anti-atoms that make up anti-matter. Just as a planet orbiting a star has the physical analog of an atom, a black hole has the physical analog of an anti-atom.

II.

Theoretical Basis for the Theory

Much of the material in this book requires some *a priori* knowledge of physics and mathematics. The reader should be conversant with basic quantum mechanics, special relativity, spacetime geometry, gravitation, the Dirac equation and spacetime algebra. This section describes to someone without a strong background in physics or mathematics the concepts behind the theory.

When Einstein came up with his theory of special relativity, he started off from the premise of light waves propagating *in vacuo* along real axes forwards in time. By considering the propagation of light waves in a flat Minkowski spacetime, the Lorentz transformation equations were derived. One frame of reference was revealed to be related to another frame of reference *via* the Lorentz factor which was shown to be positive for a propagation forwards in time.

But consider this. Consider that light waves in their own reference frame do not move forwards in time and that they do not move backwards in time. From Einstein's derivation, proper time for light waves becomes zero. Further, consider that light waves from other frames of reference move forwards in time but they also move backwards in time from other frames of reference. So *in vacuo*, a light wave propagates both forwards and backwards in time at a constant speed c.

So why has this not been observed in nature? Well what is nature? In a 'timelike' spacetime, nature is timelike. So a light wave traveling backwards in time which is 'spacelike' is not observed by us because we are not 'spacelike.' But that does not mean that it does not happen. That light waves move forwards in time as observed by 'timelike' beings but they do not move at all in time because of special relativity seems to be contradictory. But when something is 'lightlike,' it neither moves forwards nor backwards in time from its own reference frame. But it moves forwards and it moves backwards in time from other reference frames. From a timelike reference frame, something that is lightlike appears to move forwards in time. From a spacelike reference frame, something that is lightlike appears to move backwards in time. These are all consequences of special relativity. Note that the speed of light c is still the same for all inertial frames of reference. The spacelike light waves still propagate at the speed of light but they propagate in another direction, the negative time direction. All of the mathematics in this book is Lorentz covariant so the postulates of special relativity are observed.

In section IV of this book, a re-derivation of the Lorentz transformation equations is performed. In this derivation, the light waves move backwards in time along real axes and a negative Lorentz factor comes out of the derivation. A negative Lorentz factor is important because it yields a negative energy in special relativity *via* Einstein's famous equation $E = mc^2 / -\sqrt{1-(v^2/c^2)}$, where the negative Lorentz factor in the equation on the left produces negative energy. Negative momentum is also yielded and negative time because negative time is how the Lorentz transformation equations are arrived at in the derivation. Note that the negative Lorentz factor is arrived at because of the choice of a and b in the derivation. They are chosen to be symmetric to the choices for a and b in Einstein's derivation. In Einstein's derivation, $a = (\lambda + \mu)/2; b = (\lambda - \mu)/2$. For the derivation in this book, $a = (-\lambda - \mu)/2; b = (-\lambda + \mu)/2$. Symmetry plays a big role in this book.

Per special relativity, positive energy relates to moving forwards in time. Per the derivation for special relativity in this book, negative energy relates to moving backwards in time. Further, in quantum mechanics, there exist four plane wave solutions to the time-dependent Schrödinger equation. Two of these plane wave solutions solve for positive energy and two for negative energy. Plane wave solutions of the form $i(px-Et)$ and $i(-px-Et)$ are positive energy solutions and relate to Einstein's derivation of the Lorentz transformation equations for x-$ct = 0$ and $-x$-$ct = 0$. Plane wave solutions of the form $i(-px+Et)$ and $i(px+Et)$ are negative energy solutions and relate to the derivation of the Lorentz transformations in this book for $-x+ct = 0$ and $x+ct = 0$.

In section V, a 'spacelike' Minkowski spacetime is introduced to coincide with the re-derivation of the Lorentz transformation equations in section IV. The 'spacelike' Minkowski spacetime has imaginary space axes and a real time axis when reformulated for a four-dimensional Euclidean spacetime continuum. When reformulated for a four-dimensional Eucilidean spacetime continuum, the 'timelike' Minkowski spacetime has three real space axes and an imaginary time axis. So there is a direct symmetry between the 'spacelike' spacetime introduced in this book and the Minkowski spacetime that is the Euclidean geometry for special relativity.

Note that section IV talks about real space axes and section V talks about imaginary space axes. Consider the geometrical construct the pseudo-sphere. A pseudo-sphere has all of the properties of a sphere but it has an imaginary radius ir and it has a negative curvature $-1/r^2$. Likewise, a sphere has a real radius r and a positive curvature $1/r^2$. Also, the pseudo-sphere is used to describe hyperbolic geometries such as the Klein model, the Poincaré disk model, the hyperboloid model and the Minkowski projective space. So if the geometry of the 'spacelike' Minkowski spacetime that is introduced in section V is hyperbolic, then there exists a real pseudo-spherical hyperbolic geometry for it and this geometry is real for imaginary axes and an imaginary radius. So discussing imaginary axes in section V does not violate a real geometry for real axes and does not counter the solution of the Lorentz transformation equations for real axes in section IV. A pseudo-sphere has an imaginary radius but it does not exist in imaginary space. It exists in real space. So the imaginary space axes introduced in section V discuss a real pseudo-spherical spacetime.

While Einstein used a four-dimensional Euclidean continuum for special relativity, he disregarded this continuum for general relativity because a Euclidean continuum describes a *'flat'* spacetime and spacetime in general relativity is *'curved.'* So Einstein asserted that a different curved geometry needed to be used for general relativity, one based on Gaussian co-ordinates. Poincaré argued with this restriction from Einstein's equivalence principle because he believed that spacetime can be described using any set of co-ordinates, Cartesian or Gaussian. More recently, the gauge theory of gravitation has been introduced by the Cambridge group and David Hestenes[8, 9]. This theory removes the restriction of having to use a curved geometry to describe a curved spacetime. So if one prescribes to the gauge theory of gravitation as a solution for general relativity, flat Minkowski spacetime can again be used as the spacetime for general relativity.

In order to solve for general relativity as put forth by Einstein, one needs to be very conversant with complex mathematics and physics. A thorough understanding of tensors is a prerequisite. Although the primary equation for general relativity is quite elegant, the mathematics and physics of traditional general relativity is not elegant and its complexity is one of the reasons it has not been widely integrated into physics curriculum. By recasting general relativity to a flat Minkowski spacetime such as the 'timelike' spacetime discussed in section V, much of the complexity in the mathematics and physics of general relativity is removed. This is done in the gauge theory of gravity. The theory reformulates general relativity for a flat spacetime by relating general relativity to geometrical calculus which is an extension of geometric algebra or David Hestenes' spacetime algebra [STA]. It also removes the restrictions of the equivalence principle by introducing the displacement gauge principle.

So the two spacetimes discussed in section V can be used to describe flat 'light-like' spacetime or they can be used to describe curved 'timelike' spacetime or curved 'spacelike' spacetime. The restriction of the equivalence principle in general relativity forcing us to use a curved spacetime is removed in the gauge theory of gravity if one considers that global spacetime is homogeneous. In the gauge theory of gravity, by asserting the homogeneity of global spacetime, the equivalence principle is supplanted by the *'principle of gauge equivalence.'* If one considers this conceptually, it makes sense that any point in spacetime can be represented by any set of co-ordinates, Cartesian or Gaussian.

So general relativity can be solved for using a flat Minkowski spacetime. General relativity can also be solved for using the 'spacelike' spacetime that is introduced in section V. The spacelike spacetime is orthogonal to the Minkowski spacetime so the solution will be similar. The solution has not been solved for in this book but is left as a future exercise. The primary objective of this book is not to get into complicated mathematics and physics but to present these new ideas and theories and solve for some of the mathematics and physics and show how black holes and anti-matter must have an intrinsic negative spacetime curvature. This is the only way they can be properly described from the Dirac equation, quantum mechanics and relativity. This book does not solve for the anti-hydrogen atom either and this is also left as a future exercise. Also, how much anti-matter creates how much negative spacetime curvature or *vice versa* needs to be determined for compressed spacetime experimentation.

If one considers now that the spacelike spacetime introduced in section V is real and has a real pseudo-spherical, hyperbolic geometry, then let us get back to the form of these geometries. In hyperbolic geometry, the Minkowski quadratic form is defined as: $Q([x_0, x_1, ..., x_n]) = x_0^2 - x_1^2 - ... - x_n^2$. For the quadratic form, correspondingly there exists a Minkowski bilinear form as: $B(u,v) = x_0 y_0 - x_1 y_1 - ... - x_n y_n$. Consider that $n = 4$ for the four dimensions of spacetime and that the co-ordinates x_0 and y_0 are time co-ordinates and the other three co-ordinates are the space co-ordinates. The dot product between two vectors in this spacetime $B(u,v)$ is equal to $|u||v|\cosh(\theta)$ where Q is the square of the magnitude of the vectors. The angle θ is the distance in radians along a unit pseudo-sphere between the two vectors. Considering the distance as a geodesic in general relativity, there is a nice fit between the 'spacelike' spacetime introduced in section V and hyperbolic geometry. Further, while the co-ordinates for 'spacelike' spacetime are arrived at in a different more intuitive manner in section V, the co-ordinates could also have been arrived at simply by taking the square root of $Q(u,v)$. By associating the zeroeth co-ordinate with time and the other co-ordinates with space, a real time co-ordinate and imaginary space co-ordinates are produced. In this book, we can either elicit a hyperbolic geometry from a 'spacelike' spacetime as is done in section V or a 'spacelike' spacetime can be elicited from a hyperbolic geometry as we have just shown. Geometry is the true form of physics. All physical theories are tied to a geometry.

When Einstein was confronted with quantum mechanics, he said that he 'did not believe in a God that plays dice.' Was Einstein correct? In section III of this book, it is revealed that the anti-particle is a natural consequence of 'the Particle in a Box with a Finite Boundary' problem. By reformulating the problem in terms of 'the Anti-Particle in a Box,' the anti-particle is revealed. So what was originally thought to be the particle existing where it is not supposed to be because of quantum probability was really just the anti-particle in disguise. There are inherent issues with 'the Particle in a Box' problem. One of the fundamental issues is that the attenuating particle continues forever in space since it never attenuates to zero. This is related in quantum mechanics to saying that a particle can be anywhere in space from negative infinity to positive infinity. The probability is very, very small the particle will appear somewhere in space far, far away from where it is supposed to be but the probability still exists. If one believes the mathematics of quantum mechanics, then there is a one-trillionth of a trillionth of a chance that a particle will suddenly appear twenty light years away from where it is supposed to be. This does not make sense but seems to be something that is accepted in quantum mechanics. If a particle does not exist where it is not supposed to be, then perhaps what we are looking at phenomenologically is the anti-particle. In quantum tunneling, the anti-particle shows up again. The particle does not tunnel through the barrier because there is a probability it is on the other side of the barrier. The particle tunnels through the barrier because it becomes its anti-particle and the anti-particle exists on either side of the barrier.

Some more significant things come out of the 'Anti-Particle in a Box' problem that relate to all of physics. The anti-particle must have negative energy.

Because the anti-particle must have negative energy, the only plane wave solution to the time-independent Schrödinger equation for the anti-particle is a hyperbolic plane wave solution. So the 'Anti-Particle in a Box' problem links hyperbolic geometry to anti-matter. From there, all of the concepts and ideas in this book can be connected. In Section IV, the Lorentz transformation is associated to negative energy, negative momentum and moving backwards in time. In Section V, the 'spacelike' Minkowski spacetime is introduced which relates to the Lorentz transformation and to hyperbolic geometry. In section VI, the plane wave form of the positron is related to the anti-particle from section III and it is revealed that the plane wave solution has an imaginary radius for a pseudo-sphere, a hyperbolic plane waveform, a negative energy and moves backwards in time. In fact, the plane wave solution for the positron (52) introduced in section VI is the 'only' solution to the real Dirac equation that is also a solution to the time-dependent Schrödinger equation for negative energy. The form of the equation is hyperbolic just as the form for the equation in the 'Anti-Particle in a Box' problem is hyperbolic.

Mathematically and geometrically, this book makes a solid argument for anti-matter and black holes having an intrinsic negative spacetime curvature. Of course, the real beauty of the ideas and concepts presented in this book will come out when they are solved for in general relativity and for the anti-hydrogen atom. Some physical experiments to test the theories in this book are given in section VIII. The theories and ideas in this book could have been incorporated into general relativity and the anti-hydrogen atom but there is enough material compiled for an experiment to be performed to test the validity of these theories before going any further.

If black holes and anti-matter have an intrinsic negative spacetime curvature, then spacetime is compressed. Just as spacetime is expanded for matter which has gravity, spacetime is contracted for anti-matter and black holes which have anti-gravity. If this proposition is true and enough material in this book supports that it is, then maybe the human race will be able to travel through space-time. This book asks the reader to have an open mind. Why should this proposition not be true? Certainly, there is no black hole that we know of that is close enough that we can find out. But, given the nature and geometry of anti-matter presented in this book, certainly experiments can be devised to determine whether or not anti-matter acts as predicted.

Certainly, if black holes are as described in this book, they have a more elegant geometrical and mathematical form than they currently have. Inside of the event horizon, the singularity form of the black hole makes it intractable. A singularity is basically something that is hard to conceive (any number divided by zero) and at a singularity all the laws of physics break down. So using this prevailing view of a black hole inherently leads to other misconceptions. For one thing, can a quantum particle even exist in an area of spacetime that is a singularity? From quantum mechanics, probabilistically the particle can. This is counter-intuitive. The particle can probabilistically pass through the event horizon of the black hole but it cannot exist in an area of spacetime that is a singularity since all the laws of physics break down. This book recasts quantum probability as the anti-particle in disguise. Considering this, we now know that at a boundary or a barrier, the particle does not probabilistically pass through the boundary or the barrier. But at a boundary or a barrier, the particle can become its anti-particle. The anti-particle can then pass through the boundary or a barrier to exist in a 'spacelike' spacetime. At the event horizon of a black hole, the particle can become its anti-particle to enter the black hole after becoming lightlike at the event horizon. All the equations, mathematics and geometry in this book point to this as being a more rational interpretation then quantum probability. So perhaps Einstein was correct.

Certainly, this does not mean that the Heisenberg uncertainty principle should be tossed away. The Heisenberg uncertainty principle reveals a limit to spacetime. If one considers spacetime as being composed of cubes in space, then the dimension of a cube $\hbar$ will be related to the Heisenberg uncertainty principle. Certainly, these spacetime 'cubes' will be quantized with respect to wavelength, frequency and energy. A spacetime cube that has a 'timelike' Minkowski spacetime will have a positive curvature and be expanded outwards since the edges of the cube are curved outwardly. A spacetime cube that is lightlike will have flat edges and it will not be expanded nor contracted. A spacetime cube that is spacelike will have a negative curvature and be contracted since the edges of the cube are curved inwardly. So the timelike cube has a gravity corresponding to a mass m and the spacelike cube has an anti-gravity corresponding to an anti-mass m. This is all probably difficult to swallow but that does not mean that it is not true. Note that because of the gauge theory of gravitation that was discussed earlier, we can talk about the spacetime cubes for a Minkowski spacetime or for the spacelike spacetime introduced in this book. For a lightlike spacetime, both the Minkowski spacetime and the spacelike spacetime are applicable since the light waves propagate forwards and backwards along the time axis *in vacuo*. The other thing to consider is that spacetime curvature creates matter and matter creates spacetime curvature. Contracting a lightlike spacetime cube creates anti-matter. Expanding a lightlike spacetime cube creates matter. The curvature of the cube is what creates the matter or anti-matter.

III.

The Anti-Particle in a Box

Any college student who takes a beginning quantum mechanics course is probably familiar with the 'Particle in a Box with a Finite Boundary' problem. In this problem, a free particle exists in a box of length L. At the boundaries of the box, there is a potential V_0 and the particle can extend past the box into this potential region where it exponentially attenuates beneath the potential. It attenuates rapidly to C/e at $1/\alpha$ assuming a waveform

$$\Psi_I = C * \exp(\alpha x) \tag{1}$$

at the left boundary $(-L/2)$ and

$$\Psi_{III} = C * \exp(-\alpha x) \tag{2}$$

at the right boundary $(L/2)$. The waveform for the particle contained in the box is

$$\psi_{II} = B * \cos(kx). \tag{3}$$

By matching boundary conditions, the constraint equation

$$k * \tan(kL/2) = \alpha \tag{4}$$

is derived. Here the even parity solution for the 'Particle in a Box' is used. There exists an odd parity solution as well but it is not considered here.

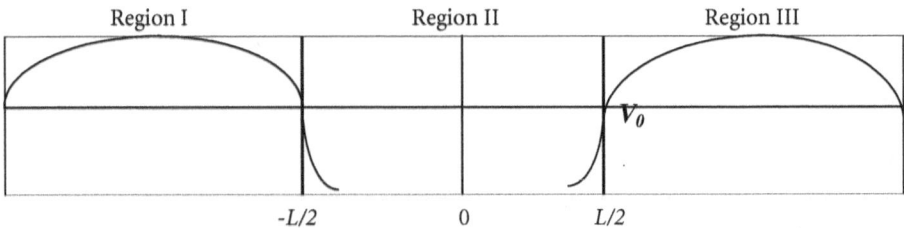

Figure 1: The Anti-Particle in a Box

Now consider the exact opposite problem as depicted in Figure 1. Consider a particle trapped beneath some potential V_0 in a box of length L. The particle has some energy $E < V_0$. At the boundaries of the box, the trapped particle has the chance of escaping its prison and becoming a free particle. So we have three regions. In Regions I and III, we have a free particle with a standard waveform of

$$\Psi_{I,III} = B * \cos (kx). \tag{5}$$

In region II, we have a trapped particle with a waveform of

$$\Psi_{II} = C * \exp (\alpha x) + C * \exp (-\alpha x). \tag{6}$$

Using the hyperbolic cosine identity $\cosh(\alpha x) = (e^{\alpha x} + e^{-\alpha x})/2$ to replace the exponential functions in region II, Ψ_{II} becomes

$$\Psi_{II} = 2C * \cosh (\alpha x). \tag{7}$$

By matching boundary conditions at $x = -L/2$ and at $x = L/2$, a symmetrical constraint equation

$$k * \tan (kL/2) = \alpha * \tanh (-\alpha L/2) \tag{8}$$

is derived. This is the same as equation (4) but with the added term $\tanh (-\alpha L/2)$.

A number of significant observations can be made:

I. The anomaly of the exponentially attenuating particle is given a physical and mathematical significance through the hyperbolic cosine plane waveform. In the 'Particle in a Box' problem, it is unclear where the attenuating portion of the particle in regions I and III attenuates to. Since it never goes to zero, it continues forever. This does not make sense. In the 'Anti-Particle in a Box' problem, the attenuating particle in region II does not continue forever because it is met at the other boundary by the other attenuating particle in region II hence forming the hyperbolic cosine plane waveform.

II. Because the waveform in region II for the 'Anti-Particle in a Box' is hyperbolic in nature, it has a negative spacetime curvature since hyperbolic geometry has negative curvature. If this problem is extended to the three dimensional 'Anti-Particle in a Cube,' the space occupied by the anti-particle is pseudo-spherically symmetrical as well.

III. Because the anti-particle attenuates to C/e after only $1/\alpha$ into the box, the length L of the box containing the anti-particle is compressed to roughly $2/\alpha$ or $2/k$. Because of the negative spacetime curvature, the space L is compressed to a much smaller value. For the 3-D 'Anti-Particle in a Cube,' the space L^3 is compressed to about $8/\alpha^3$ or $8/k^3$.

(In fact, integrating $2C^* \cosh(\alpha x)$ from $-L/2$ to $L/2$ yields the exact result $2/k$ with $k = n\pi/L$ and $C = \frac{1}{2}$ and equation (10) below.)

IV. To take this a step further, consider that the potential V_0 is actually equal to zero. Then the energy for the anti-particle is negative and the anti-particle is not trapped but is a free particle with negative energy. Then at the boundaries of the box, the anti-particle can become its inverse particle with positive energy. Let us go into more detail:

Mathematically speaking, we know from the 'Particle in a Box' problem that

$$\alpha = \frac{\sqrt{2m(V_0 - E)}}{\hbar}, k = \frac{\sqrt{2mE}}{\hbar}. \tag{9}$$

If the potential energy V_0 is zero, then α in (9) becomes

$$\alpha = \frac{\sqrt{2m(-E)}}{\hbar} = \frac{i\sqrt{2mE}}{\hbar} = ik. \tag{10}$$

Because of the hyperbolic identity $\tanh(x) = -i\tan(ix)$, the right-hand-side of equation (8) becomes

$$\alpha \tanh(-\alpha L/2) = -i\alpha \tan(-i\alpha L/2) = -i(ik)\tan(-i(ik)L/2)$$
$$= k\tan(kL/2) \tag{11}$$

which is precisely the same result as the left-hand-side of equation (8)! Is this just some strange coincidence? By assigning a hyperbolic cosine waveform solution to the 'Anti-Particle in a Box,' we have derived a constraint equation that the anti-particle has the same energy levels as the particle but that the anti-particle energy levels are all negative. Per the constraint equation (8), $E = \hbar^2 k^2 / 2m$ for the particle but $E = -\hbar^2 k^2 / 2m = \hbar^2 \alpha^2 / 2m$ for the anti-particle.

V. An anti-particle has a hyperbolic cosine waveform along the real x-axis (even parity case) or it has a normal cosine waveform along the imaginary x-axis. This is because $\cosh(x) = \cos(ix)$. Likewise, a particle has a normal waveform along the real x-axis or it has a hyperbolic cosine waveform along the imaginary x-axis. At a boundary an anti-particle can become its particle or *vice versa*.

VI. The 'Anti-Particle in a Box' problem has a direct impact on quantum tunneling. When a particle is at a barrier, it becomes its anti-particle. The anti-particle tunnels through the barrier to the other side. On the other side of the barrier, the anti-particle becomes its particle. To tunnel through the barrier, the anti-particle uses negative energy $-E$ and moves backwards through time with proper time $-\tau$. This will be explained in Section X.

IV.

The Unexpected Lorentz Transformation

And now for something completely different. By using a frame K' that moves backwards along an x-axis and a time-axis with respect to a stationary frame K, the Lorentz transformation equations are re-derived. This is important to establish Lorentz covariance for the 'spacelike' spacetime that will be described in the next section. This derivation is analogous to Einstein's for special relativity in appendix 1 of [4] for Minkowski spacetime.

Beginning with $-x = -ct$ we obtain:

$$-x + ct = 0 \qquad (12)$$

Then we have:

$$-x' + ct' = \lambda(-x + ct). \qquad (13)$$

Considering a third frame which moves along the positive x-axis, we then have:

$$x' + ct' = \mu(x + ct). \qquad (14)$$

Now adding and subtracting (13) and (14) yields:

$$-x' = ax - bct \qquad (15)$$

and

$$ct' = bx - act \tag{16}$$

where

$$a = (-\lambda - \mu)/2 \tag{17}$$

and

$$b = (-\lambda + \mu)/2. \tag{18}$$

By setting $x' = 0$ in (15), we can calculate the instantaneous velocity of K' as

$$v = dx/dt = bc/a. \tag{19}$$

By taking a 'snapshot' of K' from K and setting $t = 0$ in (15), we obtain $-x' = ax$. So

$$\Delta x = 1/(-a). \tag{20}$$

By taking a 'snapshot' of K from K' and setting $t' = 0$ in (16), we obtain

$$t = bx/ac = vx/c^2. \tag{21}$$

Subbing back in for t in (15) yields

$$-x' = ax(1 - (bc/a)(v/c^2)) = ax(1 - v^2/c^2). \tag{22}$$

So

$$\Delta x' = (-a)*(1 - v^2/c^2). \tag{23}$$

Setting Δx (20) equal to $\Delta x'$ (23) yields

$$(-a)^2 = \frac{1}{1 - \dfrac{v^2}{c^2}}; (-a) = \frac{1}{\sqrt{1 - \dfrac{v^2}{c^2}}}; a = \frac{-1}{\sqrt{1 - \dfrac{v^2}{c^2}}} \tag{24}$$

So (15) becomes

$$x' = \frac{(x - vt)}{\sqrt{1 - \dfrac{v^2}{c^2}}} \tag{25}$$

and (16) becomes

$$t' = a * ((bc/a) * (x/c^2) - t)$$

$$t' = \frac{(t - \dfrac{vx}{c^2})}{\sqrt{1 - \dfrac{v^2}{c^2}}} \tag{26}$$

Along with

$$-y' = -y; \; y' = y \tag{27}$$

and

$$-z' = -z; \; z' = z, \tag{28}$$

we have re-derived the Lorentz transformation equations but have revealed a negative Lorentz factor in (24). The reader should further consider that Einstein's derivation could have been arrived at using imaginary space and time axes. The derivation would not be any different other than the substitutions $x \to ix$, $y \to iy$, $z \to iz$ and $t \to it$, but the derivation would be analogous to the derivation in this book because moving forwards along an imaginary axis relates to moving backwards along a real axis.

To further reinforce this, if we substitute back in $-x = -ct$ into equations (25) and (26), we get $-x' = -ct'$. We know Lorentz covariance will hold true for a spacetime that is based on real axes moving forwards in time. What is important is that Lorentz covariance has been proven for an unexpected case, that of real axes moving backwards in time. This case describes another spacetime that is Lorentz covariant with respect to the Lorentz transformation which is discussed in the next section.

V.

'Spacelike' Minkowski Spacetime

Consider two equations that are significant for Minkowski spacetime and the Lorentz transformation. The first equation which characterizes a Lorentz transformation is:

$$x'^2 + y'^2 + z'^2 - c^2t'^2 = x^2 + y^2 + z^2 - c^2t^2 \tag{29}$$

If this equation is reformulated in terms of all positive generic co-ordinates $\{x_1, x_2, x_3, x_4\}$, we have

$$x_1 = x \tag{30}$$
$$x_2 = y \tag{31}$$
$$x_3 = z \tag{32}$$
$$x_4 = ict \tag{33}$$

which brings us to our next equation representing the Lorentz transformation in (29)

$$x_1'^2 + x_2'^2 + x_3'^2 + x_4'^2 = x_1^2 + x_2^2 + x_3^2 + x_4^2. \tag{34}$$

This equation has an orthonormal set of basis vectors $\{x_1, x_2, x_3, x_4\}$ which define a Lorentz covariant Minkowski spacetime. Now multiply each of these basis vectors by an imaginary number on the right-hand-side which is equiva-

lent to multiplying both sides of (29) by negative one. By doing so, we have
defined a whole new set of orthonormal basis vectors $\{x_1, x_2, x_3, x_4\}$ which are:

$$x_1 = ix \tag{35}$$
$$x_2 = iy \tag{36}$$
$$x_3 = iz \tag{37}$$
$$x_4 = -ct \tag{38}$$

This new set of orthonormal basis vectors describes a Lorentz covariant 'spacelike' spacetime as:

$$-x'^2 - y'^2 - z'^2 + c^2t'^2 =- x^2 - y^2 - z^2 + c^2t^2 \tag{39}$$

and

$$ds^2 = -x^2 - y^2 - z^2 + c^2t^2 \tag{40}$$

and

$$ds^2 = x_1^2 + x_2^2 + x_3^2 + x_4^2. \tag{41}$$

Equations (39),(40) and (41) define a 'Euclidean' four-dimensional continuum.
Equations (40) and (41) also represent the geodesic for spacelike spacetime
where (41) uses the basis set from (35-38). We will see later in this book how (40)
and (41) describe shortened spacetime distances because of particle interactions
with black holes and anti-matter. For a co-ordinate transformation from timelike spacetime to spacelike spacetime, the spacelike co-ordinate equals i times the
Minkowski co-ordinate. Likewise, the Minkowski co-ordinate equals $-i$ times the
spacelike co-ordinate. So, $x_{SL} = i * x_{TL}$ and $x_{TL} = -i * x_{SL}$.

Note how (39) and (29) relate to each other. All of the space co-ordinates for
(39) are imaginary and the real time co-ordinate for the 'spacelike' spacetime is
the opposite of the imaginary time co-ordinate for the Minkowski spacetime.
The real and imaginary features of the spacetime have been flipped to create a

'spacelike' spacetime. Because all of the space co-ordinates are imaginary, the 'spacelike' spacetime is a space that is pseudo-spherically symmetrical and hyperbolic in nature and has a negative spacetime curvature. The 'spacelike' and Minkowski spacetimes are orthogonal to each other where the 'spacelike' spacetime is the imaginary part for the real spacetime. So …

$$M^4 = \text{Re}\{M^4\} + \text{Im}\{M^4\} \tag{42}$$

Also note how equations (35-38), (40) and (41) can be used to describe a pseudo-spherical hyperbolic geometry such as the Klein model, the Poincaré disk model, the hyperboloid model or the Minkowski projective space. Consider a distance for a unit pseudo-sphere measured in radians between two vectors u and v in these geometries as:

$$d(u,v) = \text{arccos}\, h(\frac{B(u,v)}{\sqrt{Q(u)Q(v)}}) \tag{43}$$

Notice that $B(u,v)$ is just the dot product of two vectors u and v that use the basis set from (35-38) and $Q(u)$ and $Q(v)$ are equivalent to the geodesic in (40). So there is a harmonic fit between the 'spacelike' spacetime defined by (35-38),(39), (40) and (41) and existing pseudo-spherical hyperbolic geometries. Further consider that the distance for a unit sphere measured in radians between two vectors u and v for Minkowski spacetime is:

$$d(u,v) = \text{arccos}(\frac{B(u,v)}{\sqrt{Q(u)Q(v)}}) \tag{44}$$

Here we have two vectors u and v defined on a Minkowski spacetime using the basis set from (30-33). The symmetry between (43) and (44) is elegant. Further, (44) describes the distance traveled along a sphere with a real radius r and (43) describes the distance traveled along a pseudo-sphere with an imaginary radius ir. (For a unit sphere, $r = 1$.)

Let us simplify this a bit more. If we say that u and v are unit vectors and consider motion only along the x-axis, then we have a 'timelike' length of traversal of:

$$L_{TL} = \arccos(x_1 x_2 - c^2 t_1 t_2). \tag{45}$$

Now if we perform a co-ordinate transformation from timelike co-ordinates to spacelike co-ordinates in (45), we have:

$$
\begin{aligned}
x &\to ix \\
t &\to it \\
L &\to iL
\end{aligned}
\tag{46}
$$

which yields:

$$L_{SL} = \arccos h(c^2 t_1 t_2 - x_1 x_2) \tag{47}$$

which is the spacelike length of traversal.

VI.

The Real Dirac Equation

The relativistic classical Dirac equation for a single particle electron predicts four different states for the particle because of the four Dirac matrices. Two of these states have positive energy E corresponding to two different spin states for the electron (spin up and spin down) and the other two have negative energy $-E$ corresponding to two different spin states for the positron (spin up and spin down.) By reformulating the Dirac equation in terms of a Spacetime Algebra [STA] or Clifford Algebra, David Hestenes came up with the real Dirac equation. The real Dirac equation eliminates imaginary numbers from the equation by replacing them with a geometrical inner product but for more information on the real Dirac equation refer to David Hestenes' seminal papers [1][2]. Anyways, by reformulating the classical equation in terms of a different algebra, some new things come out of the equation. First, consider David Hestenes' equations that he comes up with as positive energy plane wave solutions to the real Dirac equation for the electron and the positron.

His plane wave solution for the electron (for one spin state) is:

$$\Psi_- = \sqrt{\rho} R_0 e^{-ip \bullet x/\hbar} \tag{48}$$

His plane wave solution for the positron (for one spin state) is:

$$\Psi_+ = \sqrt{\rho} iR_0 e^{+ip \bullet x/\hbar}$$

(49)

The problem with (49) is that it is not a solution of the time-independent Schrödinger equation (54) for a negative energy state. From the classical Dirac equation, we know that positrons have negative energy. Also, in David Hestenes' paper [1], negative energy is predicted by the equation

$$p = mve^{-i\beta}$$

(50)

If β is equivalent to 180 degrees, then

$$p = -mv$$

(51)

corresponding to the positron. By enforcing two constraints, we come up with an alternative plane wave solution for the positron to the real Dirac equation. First, the space part of the solution must be hyperbolic. Second, the solution must be a negative energy solution to the time-independent and time-dependent Schrödinger equations. By enforcing these two conditions, we come up with the plane wave solution for the positron as:

$$\Psi_+ = \sqrt{\rho}(iR_0)e^{\pm \alpha x + iEt/\hbar}$$

(52)

where $\alpha = ik$ as in section III equation (10). Note that if we make the substitution $\alpha = ik$ in equation (49), equations (49) and the $-\alpha$ form of equation (52) are equivalent. So while (49) is a solution to the time-dependent Schrödinger equation for a negative energy solution, it is not a solution to the time-independent Schrödinger equation where (52) is a solution to both. Here we use the simple 1-D version of the time-dependent Schrödinger equation:

$$\frac{-\hbar^2}{2m}\frac{d^2\Psi}{dx^2} = i\hbar\frac{d\Psi}{dt}$$

(53)

and the time-independent Schrödinger equation:

$$\frac{-\hbar^2}{2m}\frac{d^2\Psi}{dx^2} = E\Psi.$$

(54)

Why is (52) significant? If we look at the waveform for (48), we have some-thing that is sinusoidal and spherically symmetrical. The waveform for (49) is sinusoidal but it is pseudo-spherically symmetrical because the radius is imag-inary: iR_0 (relating i geometrically to R_0.) But (52) is hyperbolic because the i for the space part has been eliminated from the superscript of e and it is pseudo-spherically symmetrical because of its radius $R = iR_0$. The importance of being both pseudo-spherically symmetrical and hyperbolic is that these are both characteristics of a spacelike spacetime that has a negative curvature. A hyperbolic geometry has negative curvature. A pseudo-sphere has negative curvature. The real question is 'does a positron have positive energy or negative energy?' It has negative energy.

Geometrically relating i to R_0 is reinforced because (52) is 'spacelike.' If (52) is 'timelike,' then the radius R will take the form:

$$R = \sqrt{x^2 + y^2 + z^2}.$$

(55)

However, (52) is 'spacelike,' so the radius R takes the form:

$$R = \sqrt{(ix)^2 + (iy)^2 + (iz)^2} = i\sqrt{x^2 + y^2 + z^2}.$$

(56)

This leads to a clearer geometrical and conceptual interpretation of a pseudo-spherical hyperbolic geometry such as that referred to in section V. An imagi-nary radius has a direct physical interpretation but a negative radius does not just as imaginary time cannot be conceptualized but negative time can.

If (54) describes negative energy, then plane wave solutions for the anti-particle are going to be of the form $A^*sinh(\alpha x)$ and $B^*cosh(\alpha x)$ or of the form (52). If (54) describes positive energy, then plane wave solutions are going to of the form $A^*sin(kx)$ and $B^*cos(kx)$ or of the form (49). Of course $\alpha = ik$ for a negative energy solution as seen in section III, but what is important is we have a hyperbolic negative energy solution to the time-independent Schrödinger equation (54) for some as yet undetermined constant α. This is because $-\hbar^2/2m * (ik) * (ik)$ is positive for positive energy in (54) but $-\hbar^2/2m * (\alpha) * (\alpha)$ is negative for negative energy in (54).

Of course, $-\hbar^2/2m * (\alpha) * (\alpha)$ is positive for (53) if we consider that $\alpha = ik$ but for negative energy this requires a negative right-hand-side of the time-dependent (53) which both (49) and (52) yield. Hence both (49) and (52) are solutions to the time-dependent form of the Schrödinger equation for negative energy which also questions the validity of claiming that (49) is a positive energy solution. Only waveform equations of the form $i(px - Et)$ and $i(-px - Et)$ are positive energy solutions to the time-dependent equation. Equations of the form $i(px + Et)$ and $i(-px + Et)$ are negative energy solutions. Equation (49) is of the form $i(-px+Et)$ so it is a negative energy solution. This further reinforces the mathematical form of (52) as being correct since if (49) is a negative energy solution to (53) then it must be a negative energy solution to (54) as well and the only way that is possible is if (52) is used. The plane waveform solutions are important as well because they relate to the Lorentz transformation from section IV. The plane waveforms $px-Et$ and $-px-Et$ relate to Einstein's derivation. The plane waveforms $px+Et$ and $-px+Et$ relate to the derivation in this book which reveals a negative Lorentz factor for a spacelike spacetime.

From quantum mechanics, we have further that the kinetic energy $E = p^2/2m$. This presents a problem if the momentum is negative and the energy is negative as well. The only way to resolve that is to say that the momentum is imaginary or that the mass m is negative or if we consider the alternative form $E = \frac{1}{2}(-mv)v$ for the kinetic energy and the momentum is as in equation (51). What this reveals is that sometimes a different form or a reinterpretation of the same equation can reveal something about the physics that is hidden. Similarly, equation (52) is just a different form of (49) but new physics is revealed by the structure of (52). The new physics is that the equation describes a hyperbolic and pseudo-spherical geometry for a negative energy solution.

From the 'Anti-Particle in a Box' problem in section III, negative energy is required to change a particle into its anti-particle state for a potential energy of zero and for that anti-particle then to exponentially attenuate into a region it should not be in. So in section III we see this negative energy for the anti-particle and we see this anti-particle take on a hyperbolic waveform. Similarly, (54) describes a negative energy for the anti-particle and a positive energy for the particle and since (52) is both pseudo-spherically symmetrical and hyperbolic in space, it describes the correct geometry for the positron and is the correct waveform.

The Mysterious Parameter 'β'

One of the most significant discoveries of David Hestenes' real Dirac equation is the mysterious parameter 'β' [2]. We have already seen this parameter in (50). When β is zero degrees, we have a positive momentum mv and positive energy corresponding to an electron. When β is 180 degrees, we have a negative momentum $-mv$ and negative energy corresponding to a positron. β also partially determines the form of the plane wave solution to the real Dirac equation because part of the solution is $e^{i\beta/2}$. When β is zero, the plane wave solution contribution is 1. When β is 180 degrees, the plane wave solution contribution is i and we see this contribution in equations (49) and (52). This would suggest that at the very least, β differentiates between the plane wave states of an electron and a positron. The theory about the true physical nature of the parameter 'β' is now given. There may be some redundancy in the following material but it is important to show how β relates to state changes in different contexts. There are three different contexts. First we just describe what happens to a particle as the parameter changes. Next, we discuss how the Minkowski light cone relates to the state changes and a spacetime circle is defined that is super-imposed on the light cone. Lastly in section VII, we describe how the parameter changes relate to a particle traversing through a black hole.

Proposition 1: 'β' is a parameter that differentiates between the 'timelike', 'lightlike' and 'spacelike' states of a particle.

When β is zero degrees, we have a 'timelike' particle moving with proper time τ. When β moves from zero to ninety degrees, the particle is still 'timelike' and accelerating towards the speed of light. When β is ninety degrees, the particle is 'lightlike' and its mass has been completely converted into energy. When β

moves from ninety degrees to 180 degrees, the anti-particle is 'spacelike' and being decelerated from the speed of light to an anti-particle moving with proper time -τ at 180 degrees. When β moves from 180 degrees to 270 degrees, the anti-particle is 'spacelike' and being accelerated towards the speed of light. At 270 degrees, the anti-particle again becomes 'lightlike' and the anti-mass is exchanged for energy. From β equals 270 degrees to 360 degrees, the particle is 'timelike' again and decelerating from the speed of light back to moving at its proper time τ.

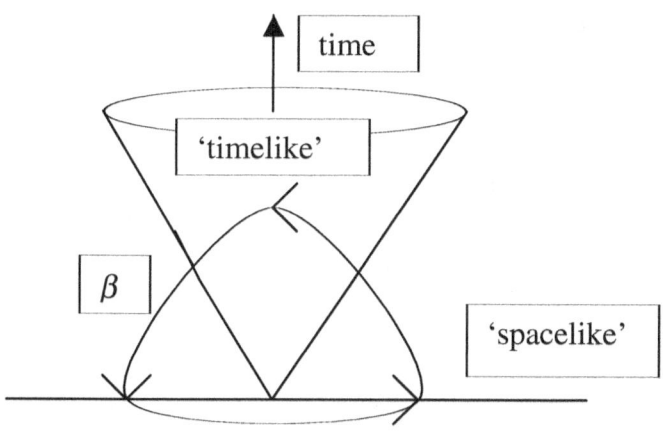

'hypersurface of the present'

Figure 2: The Minkowski light cone super-imposed with the parameter β

Consider a light cone in Minkowski spacetime as in figure 2. Any point within the future light cone is going to coincide with a real radius *r*. Any point outside of the future light cone is going to have an imaginary radius *ir*. So anything within the future light cone is going to be spherically symmetrical while anything outside of it is going to be pseudo-spherically symmetrical. When the particle moves from being 'timelike' to 'lightlike' to 'spacelike,' the radius becomes imaginary and stays that way until the particle moves back from

'spacelike' to 'lightlike' to 'timelike.' Of course, when the particle is 'lightlike' it
does not have a radius because the spacetime is flat.

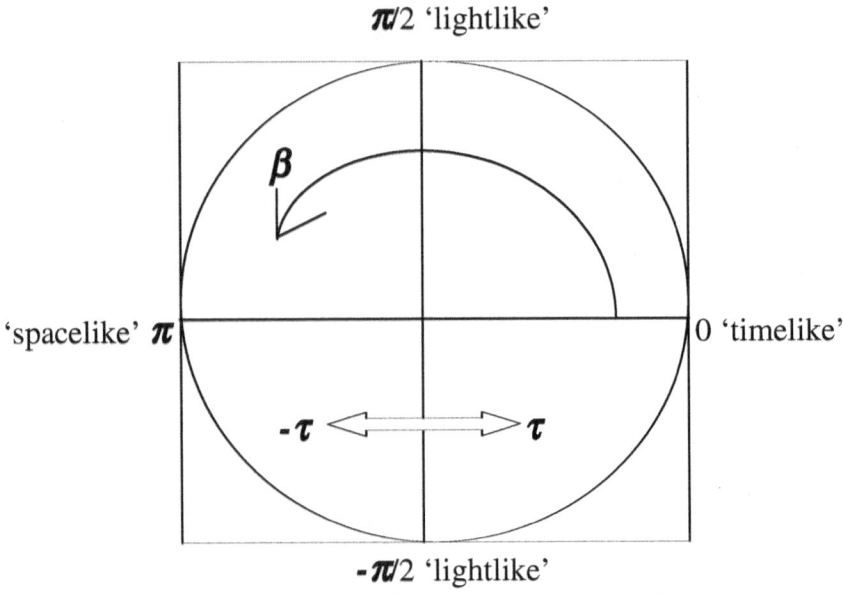

Figure 3: The Spacetime Circle

Now consider a circle that describes the movement of the parameter 'β' from
zero degrees to 360 degrees as in figure 3. Zero degrees will correspond to the
positive x-axis of the circle and ninety degrees to the positive y-axis. Now align
the positive x-axis of this circle with the positive time axis of the light cone
from figure 2 and the positive y-axis with a vector along the future light cone at
45 degrees and center the origin of the circle with the origin of the light cone.
Now the circle that describes β describes the changing of the state of the particle in the spacetime for the light cone. As β goes from zero to ninety degrees,
the particle moves from the positive time axis to the future light cone and
becomes 'lightlike.' As β goes from ninety degrees to 180 degrees, the anti-particle moves from the future light cone to the '*hypersurface of the present.*' As β

moves from 180 degrees to 270 degrees, the anti-particle moves from the *'hypersurface of the present'* back to the future light cone where it becomes 'lightlike' again. From 270 degrees to 360 degrees, the particle moves from the future light cone back to the positive time axis. Note that while the particle took on different states, it never moved 'backwards in time.' It was always at or above the *'hypersurface of the present.'*

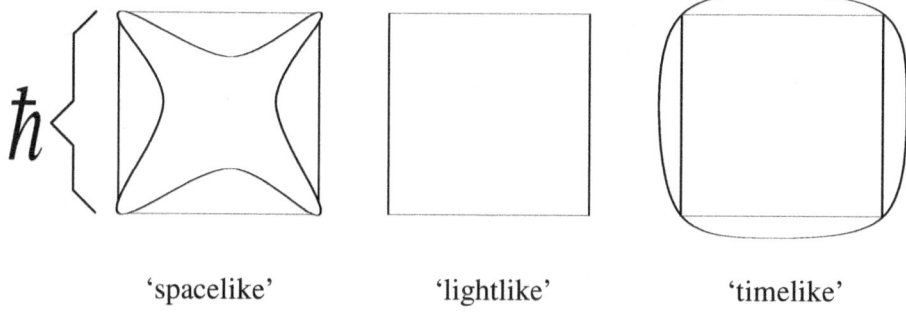

'spacelike'   'lightlike'   'timelike'

Figure 4: Spacetime squares. These can all be extended to spacetime cubes.

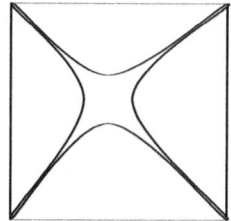

Figure 5: High energy spacelike square

VII.

The Motion of a Particle through a Black Hole

So what have we done? Why is this circle that describes the sweeping out of the parameter β physically significant?

Proposition II: Sweeping the parameter β from zero degrees to 360 degrees describes the state changes for a particle as it approaches, enters and leaves a black hole.
Proposition III: At the event horizon of a black hole, the black hole becomes 'lightlike.'
Proposition IV: Inside the event horizon of the black hole, the black hole is pseudo-spherically symmetrical, hyperbolic and 'spacelike' corresponding to a negative spacetime curvature.

Consider a particle floating through spacetime approaching a black hole. The particle is initially stationary with its proper time τ. As β moves from zero to ninety degrees, the particle is accelerated from rest towards the speed of light at the event horizon. At ninety degrees, the particle reaches the event horizon and becomes 'lightlike.' From β equals ninety degrees to 180 degrees the anti-particle is decelerated from the speed of light to its proper time $-\tau$. At 180 degrees, the anti-particle is at the center of the black hole in the 'eye of the storm.' From 180 degrees to 270 degrees, the anti-particle is accelerated back towards the speed of light and towards the event horizon on the opposite side of the black hole from which it entered. At 270 degrees, the anti-particle becomes 'lightlike' again at the event horizon. Past 270 degrees, the particle decelerates from the speed of light as it leaves the black hole and is again at rest with its proper time τ.

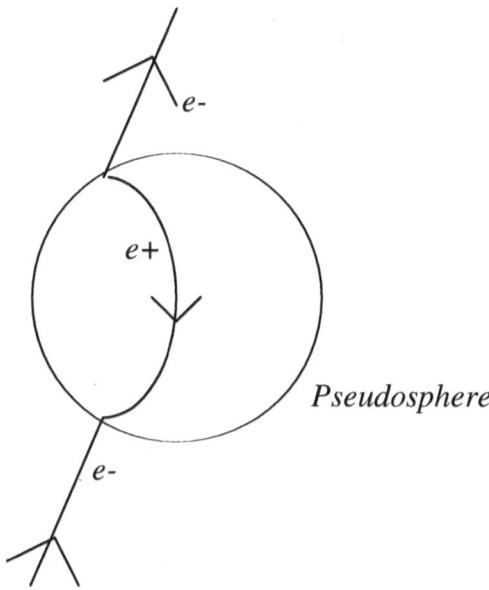

Figure 6: The motion of an electron as it approaches, enters and exits a black hole. Before entering and after exiting the black hole, the velocity of the electron is unchanged in both magnitude and direction (before being accelerated and after being decelerated from the speed of light that is.) Bear in mind that the electron is moving at the speed of light once it reaches the event horizon and upon exiting the black hole.

Let us give this particle an initial velocity v_0 before interacting with the black hole. At a point a in spacetime, the particle begins to be accelerated towards the black hole. At a point b in spacetime on the other side of the black hole, the particle will no longer be interacting with the black hole and will again have its initial velocity of v_0 (given that there is always '*some*' interaction between two gravitational bodies but the interaction is not significant.) Now let us consider an identical particle with a velocity v_0 that moves from spacetime point **a** to spacetime point **b** but does not interact with a black hole. The distance that the particle that does not interact with the black hole travels from point **a** to point **b** is going to be much longer than the distance traveled by the particle that does interact with the black hole.

Proposition V: Black Holes and Anti-Matter shorten real spacetime distances.

VIII.

The Freeman Drive

In accordance with proposition V, the concept of the freeman drive is now presented. If we consider the preceding paragraph, if the black hole were to be moving at a speed v_0 in the same direction as the particle and the center of the black hole were to be placed just in front of the particle, the particle could travel from one place to another in a lot less time than if the black hole were not present.

If the particle is replaced by human beings in a container and the black hole is replaced by a chunk of anti-matter, than we would have something that is important for travel. I call this invention 'The Freeman Drive.' The container is any kind of passenger vehicle and the engine for this vehicle (like the engine for a car) is a chunk of anti-matter that is not stationary and moving *'just in front'* of the container at the same speed and in the same direction as the container and its passengers.

One way to consider the feasibility of 'the Freeman Drive' and test the theories in this book is to pass a photon through a stationary anti-hydrogen atom and see how it is affected. From the theories in this book, the photon will enter and exit the anti-hydrogen atom unscathed but will arrive at its destination in a shorter time than if the anti-hydrogen atom were not present. This brings up the issue of how much anti-matter creates how much negative spacetime curvature or how much negative spacetime curvature creates how much anti-matter since the two are intrinsic to each other. There would have to be enough anti-matter that the arrival time for the photon is measurably smaller. What quantity this is has not yet been determined but can be calculated using the equations presented in this book. Another experiment is to measure the amount the photon is deflected and it is exactly the amount from the experiments that have been performed before for general relativity but in the opposite direction. So the photon is not '*bent in*' by the gravitational pull but it is '*bent out.*'

IX.

Mathematical Form of β

An ontological meaning has been attributed to β but it needs to be given some mathematical formalism. From special relativity, proper time τ is related to time t by:

$$\tau = t * \sqrt{1 - \frac{v^2}{c^2}} . \tag{57}$$

For the parameter β, τ is going to be positive from zero to ninety degrees and from 270 to 360 degrees and τ is going to be negative from 90 to 270 degrees. So $\tau = t * \cos(\beta)$. Relating $\cos(\beta)$ to (57) gives:

$$\cos(\beta) = \sqrt{1 - \frac{v^2}{c^2}} . \tag{58}$$

Relating spacelike proper time τ_{SL} to timelike t by using the Lorentz factor from (24) in section IV,

$$\tau_{SL} = -t * \sqrt{1 - \frac{v^2}{c^2}} . \tag{59}$$

So,

$$\cos(\beta) = \pm\sqrt{1 - \frac{v^2}{c^2}}.$$

(60)

depending on whether the proper time is spacelike or timelike. Using the standard trigonometric identity $\sin^2(\theta) + \cos^2(\theta) = 1$, we have

$$\sin(\beta) = \frac{v}{c}$$

(61)

and

$$\tan(\beta) = \frac{v/c}{\pm\sqrt{1 - \frac{v^2}{c^2}}}$$

(62)

Further, a rotation R for David Hestenes' spacetime algebra (STA) can be defined as:

$$R = e^{i\beta} = \cos(\beta) + i\sin(\beta) = \pm\sqrt{1 - \frac{v^2}{c^2}} + iv/c.$$

(63)

and a rotation $\tilde{R}$ can be defined as:

$$\tilde{R} = e^{-i\beta} = \cos(\beta) - i\sin(\beta) = \pm\sqrt{1 - \frac{v^2}{c^2}} - iv/c$$

(64)

where. $R\tilde{R} = \tilde{R}R = 1$

If β is 180 degrees for the positron and we make the substitution $\gamma = \beta/2 =$ ninety degrees in equation (63), then (63) $= i$ and we have the positron plane wave contribution to (49) and (52) from section VI.

Also, from David Hestenes' paper, we can now give an alternative physical definition to the divergence equation for the vector spin density as:

$$\partial \bullet (\rho s) = -m\rho\sin(\beta) = -m\rho v/c.$$

(65)

I am not sure what the physical significance of (65) is but perhaps the alternative form of the equation will reveal something about its physics. We now know that the divergence of the spin density when the particle is 'lightlike' is $-m\rho$ since $v = c$.

X.

Backwards in Time?

An anti-particle moving backwards in time with a proper time -τ is elicited from the time-dependent Schrödinger equation (53) and from the real Dirac equation and negative energy. This is because of the term $+iEt/\hbar$ in the plane waveform solution for the positron (52). If a particle moves forwards in time *via* $-iEt/\hbar$, then an anti-particle moves backwards in time *via* $+iEt/\hbar$.

A particle and its anti-particle are not two separate entities. An anti-particle and a particle are two different states of the same entity. The Dirac equation for a single particle electron reveals four different states for that electron. It does not reveal four different particles. From particle physics, it may *'seem like'* the electron and the positron are two different particles. But they are not. So we can say that a particle and its anti-particle are one and the same. An anti-particle moving backwards in time with a proper time -τ is the same as its particle not moving forwards in time. A particle moving forwards in time with a proper time τ is the same as its anti-particle not moving backwards in time. So when the anti-particle becomes active, the particle becomes inactive and *vice versa*. So when the anti-particle has a negative energy $-E$, its particle has an energy of zero and when the particle has an energy E, its anti-particle has a negative energy of zero. Note that this is not necessarily absolute. If a particle has some energy $E' < E$, then its anti-particle has negative energy $E' - E$ but this may not be observed in nature because energy is quantized.

Proposition VI: A particle and its anti-particle are different states of the same entity. Because of this, energy E and proper time τ are exchanged between the particle and its anti-particle depending on the state of the entity.

Proposition VII: An anti-particle moving forwards in time along its imaginary time axis is equivalent to that anti-particle moving backwards in time along its real time axis.

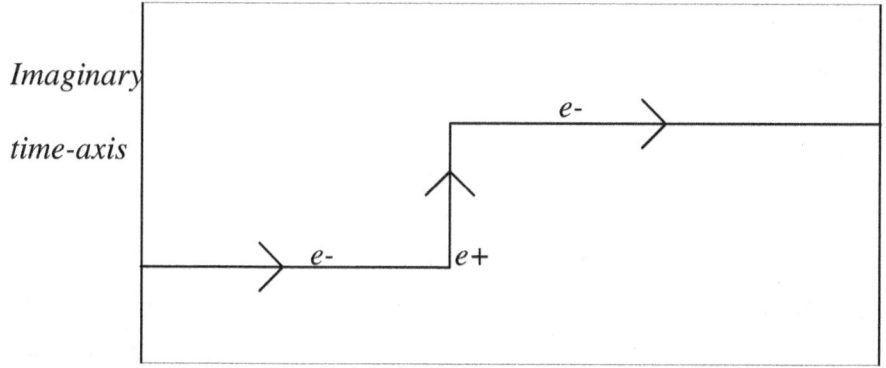

Figure 7: Diagram showing an electron becoming a positron and then becoming an electron again. While the electron is a positron it suspends the electron from moving forwards along the time axis. Note that the positron moves forwards along the imaginary time axis which is orthogonal to the real time axis.

XI.

The Anti-Hydrogen Atom

As was mentioned before in this book, the anti-atom is a physical analog for a black hole. Consider the anti-hydrogen atom. The anti-hydrogen atom has a positively charged positron that orbits a negatively charged nucleus. If we use the plane wave equation for the positron (52) and the Dirac equation to solve for a Coulomb potential, we can come up with the equations for the anti-hydrogen atom. While that has not been done here, this book proposes that the form of the anti-hydrogen atom will have the same form as a black hole, that it will have an imaginary radius ir, be pseudo-spherically symmetrical and have a hyperbolic geometry and a negative spacetime curvature. The anti-hydrogen atom is a miniature black hole.

XII.

General Relativity

For the time being, let us not consider the cosmological constant or the big bang theory and focus on the physics of general relativity as it relates spacetime curvature to energy. It is left as a future exercise to incorporate CSA into general relativity. General relativity stems from the concept of a 'timelike' geodesic taking on the form and using the notion from GTG that flat Minkowski spacetime can be used to describe curvature in GR:

$$ds^2 = dx^2 + dy^2 + dz^2 - c^2 dt^2. \tag{66}$$

This 'timelike' geodesic has a positive spacetime curvature and positive energy as it relates to the stress-energy tensor.

Now consider the 'spacelike' geodesic:

$$ds^2 = -dx^2 - dy^2 - dz^2 + c^2 dt^2. \tag{67}$$

From CSA, relating the 'spacelike' geodesic in (67) to the stress-energy tensor in general relativity will reveal a negative spacetime curvature and a negative energy.

XIII.

Conclusion

In no way is this book trying to suggest that one can move backwards in time or that an H.G. Wells time machine can be constructed. Traveling backwards in time is paradoxical. That is why in section VI it was pointed out that a particle moving through all angles of state changes in spacetime never in fact goes into the past. It never goes below the plane in the light cone that defines the '*hypersurface of the present*'. In section X, it was further explained how an anti-particle moves backwards in time and how that does not imply that particles move backwards in time. An anti-particle moving backwards in time is the same as a particle not moving forwards in time or moving along the '*hypersurface of the present*' from the Minkowski light cone. Time can be suspended for the particle but it cannot be reversed.

The most salient feature of this book is that it gives an ontological meaning and shows mathematically how compressed spacetime, anti-matter and black holes are real as they relate to quantum mechanics, the Dirac equation, Lorentz covariance and relativity. Black holes (past the event horizon) and anti-matter have these properties:

1. a 'spacelike' Minkowski spacetime

2. they are Lorentz covariant with respect to the Lorentz transformation

3. an imaginary radius *ir*

4. a negative spacetime curvature

5. they are pseudo-spherically symmetrical

6. a hyperbolic geometry

7. a negative energy *–E*

8. a negative proper time -τ

and last but not least …

9. they shorten real spacetime distances

The most important point this book makes is that a particle and its anti-particle are different states of the same entity. This book also gives an ontological meaning and mathematically solves for the mysterious parameter 'β' that is elicited from the real Dirac equation.

Physics is most beautiful when it is simple and elegant. This book takes a phenomena that is viewed as a 'cosmological dumping ground' and gives it a geometry that is more easily digestable. The solutions for anti-matter primarily stem from original thoughts towards black holes. Everything in nature that is cosmological in scale has a smaller analog. The planet orbiting a star has the analog of an atom. For the black hole, its analog is the anti-hydrogen atom.

XIV.

References

Most of the reference material used to compile this book was gleaned from the internet at the free internet encyclopedia http://www.wikipedia.org. A complete list of the topics that were used from wikipedia is:

- 'Finite potential well' for the 'Anti-particle in a box' in section III
- 'Quantum tunneling'
- 'Hyperbolic function'
- 'Hyperbolic geometry'
- 'Pseudosphere'
- 'Hawking radiation'
- 'Black hole thermodynamics'
- 'Lorentz transformation'
- 'Proper time'
- 'Minkowski space'
- 'Spacetime'
- 'Light cone'
- 'Minkowski diagram'

- 'Lorentz covariance'

- 'Time dilation'

- 'Special relativity'

- 'General relativity'

- 'Schwarzschild metric'

- 'Schwarzschild radius'

- 'Particle in a box'

- 'Schrödinger equation'

- 'Dirac equation'

- 'Complex number'

- 'Poincaré half plane model'

- 'Poincaré disk model'

- 'Klein model'

- 'Hyperboloid model'

To look up any of these topics on wikipedia, go to http://en.wikipedia.org/wiki/ and append the topic name as listed above with any spaces replaced by an '_' character. Also from the internet, a geometric algebra tutorial was used from http://www.mrao.cam.ac.uk/~clifford/introduction/intro/intro.html and a general relativity tutorial was used from http://math.ucr.edu/home/baez/gr/. While referenced below as well, David Hestenes' papers can also be found at: http://modelingnts.la.asu.edu/. The other references used to compile this book are:

[1] D. Hestenes, "Spacetime Physics with Geometric Algebra," Am. J. Phys. **71** 691-714 (2003).

[2] D. Hestenes, "Mysteries and Insights of Dirac Theory," Annales de la Fondation Louis de Broglie **28** 390-408 (2003).

[3] D. Hestenes, "Oersted Medal Lecture 2002: Reforming the Mathematical Language of Physics," Am. J. Phys. **71** 104–121 (2003).

[4] A. Einstein, Relativity, The Special and the General Theory, (Penguin Books, London, 2006).

[5] A. Einstein, Einstein's 1912 Manuscript on the Special Theory of Relativity, (George Braziller, New York, 2004).

[6] S. Hawking and R. Penrose, The Nature of Space and Time, (Princeton University Press, Princeton, New Jersey, 2000).

[7] G. Uhlenbeck & O. Laporte, "New Covariant Relations Following from the Dirac Equations," Phys. Rev. **37** 1552–1554 (1931).

[8] D. Hestenes, "Gauge Theory Gravity with Geometric Calculus," Foundations of Physics, **35(6)**: 903-970 (2005).

[9] A. Lasenby, C. Doran and S. Gull, "Gravity, Gauge Theories and Geometric Algebra," Phil. Trans. Roy. Soc. Lond. A, **356**: 487-582 (1998).

978-0-595-43093-2
0-595-43093-7